AF228581

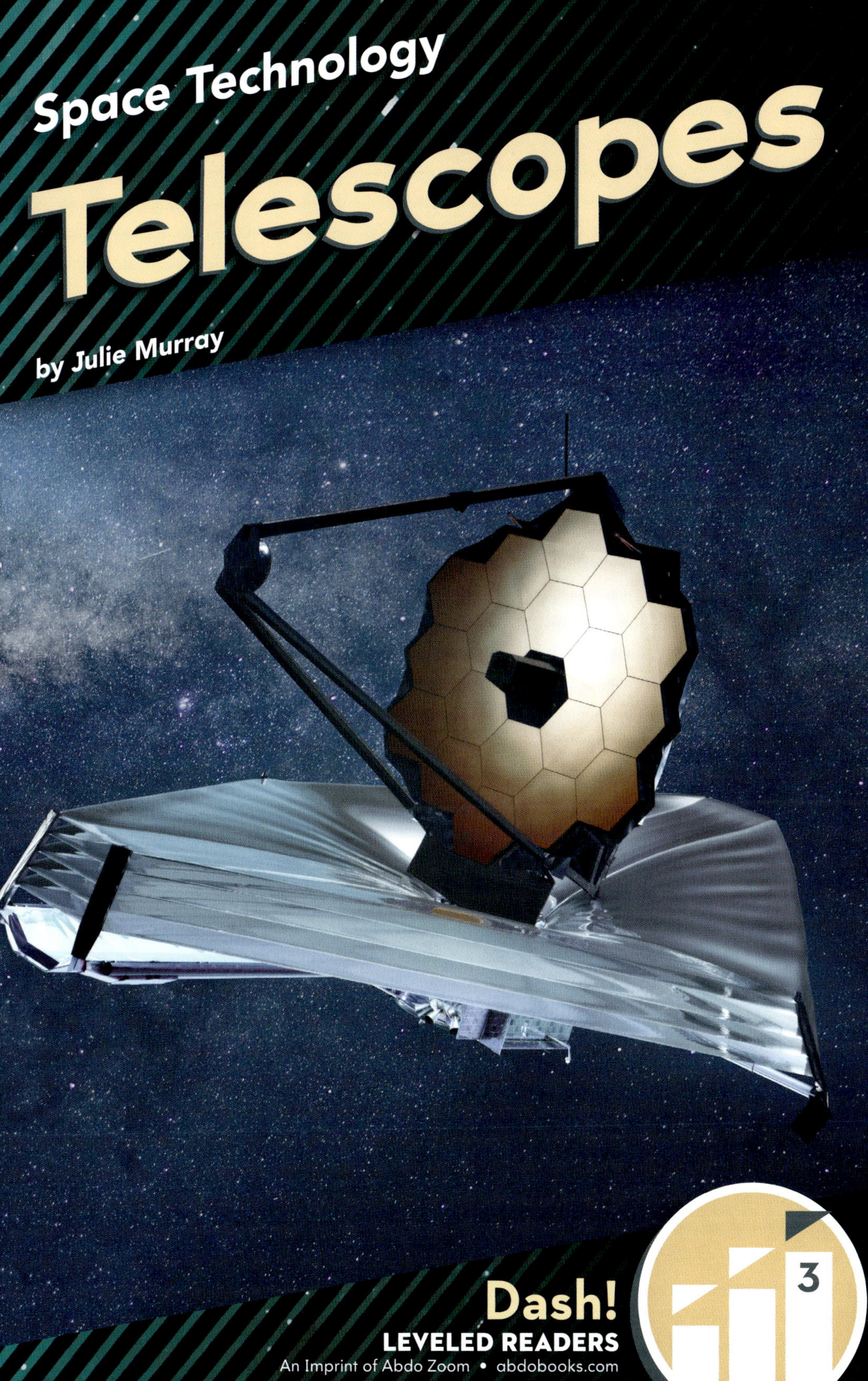
Space Technology
Telescopes
by Julie Murray
Dash!
LEVELED READERS
An Imprint of Abdo Zoom • abdobooks.com
3

Level 1 – Beginning
Short and simple sentences with familiar words or patterns for children who are beginning to understand how letters and sounds go together.

Level 2 – Emerging
Longer words and sentences with more complex language patterns for readers who are practicing common words and letter sounds.

Level 3 – Transitional
More developed language and vocabulary for readers who are becoming more independent.

abdobooks.com

Published by Abdo Zoom, a division of ABDO, PO Box 398166, Minneapolis, Minnesota 55439. Copyright © 2020 by Abdo Consulting Group, Inc. International copyrights reserved in all countries. No part of this book may be reproduced in any form without written permission from the publisher. Dash!™ is a trademark and logo of Abdo Zoom.

Printed in the United States of America, North Mankato, Minnesota.
102019
012020

Photo Credits: Alamy, Getty Images, iStock, NASA, Shutterstock, NASA/Desiree Stover
Production Contributors: Kenny Abdo, Jennie Forsberg, Grace Hansen, John Hansen
Design Contributors: Dorothy Toth, Neil Klinepier, Victoria Bates

Library of Congress Control Number: 2019941222

Publisher's Cataloging in Publication Data

Names: Murray, Julie, author.
Title: Telescopes / by Julie Murray
Description: Minneapolis, Minnesota : Abdo Zoom, 2020 | Series: Space technology | Includes online resources and index.
Identifiers: ISBN 9781532129315 (lib. bdg.) | ISBN 9781098220297 (ebook) | ISBN 9781098220785 (Read-to-Me ebook)
Subjects: LCSH: Telescopes--Juvenile literature. | Stargazing--Juvenile literature. | Space sciences--Juvenile literature. | Technology--Juvenile literature. | Stars--Juvenile literature. | Astronautics--Juvenile literature.
Classification: DDC 522--dc23

Table of Contents

Telescopes

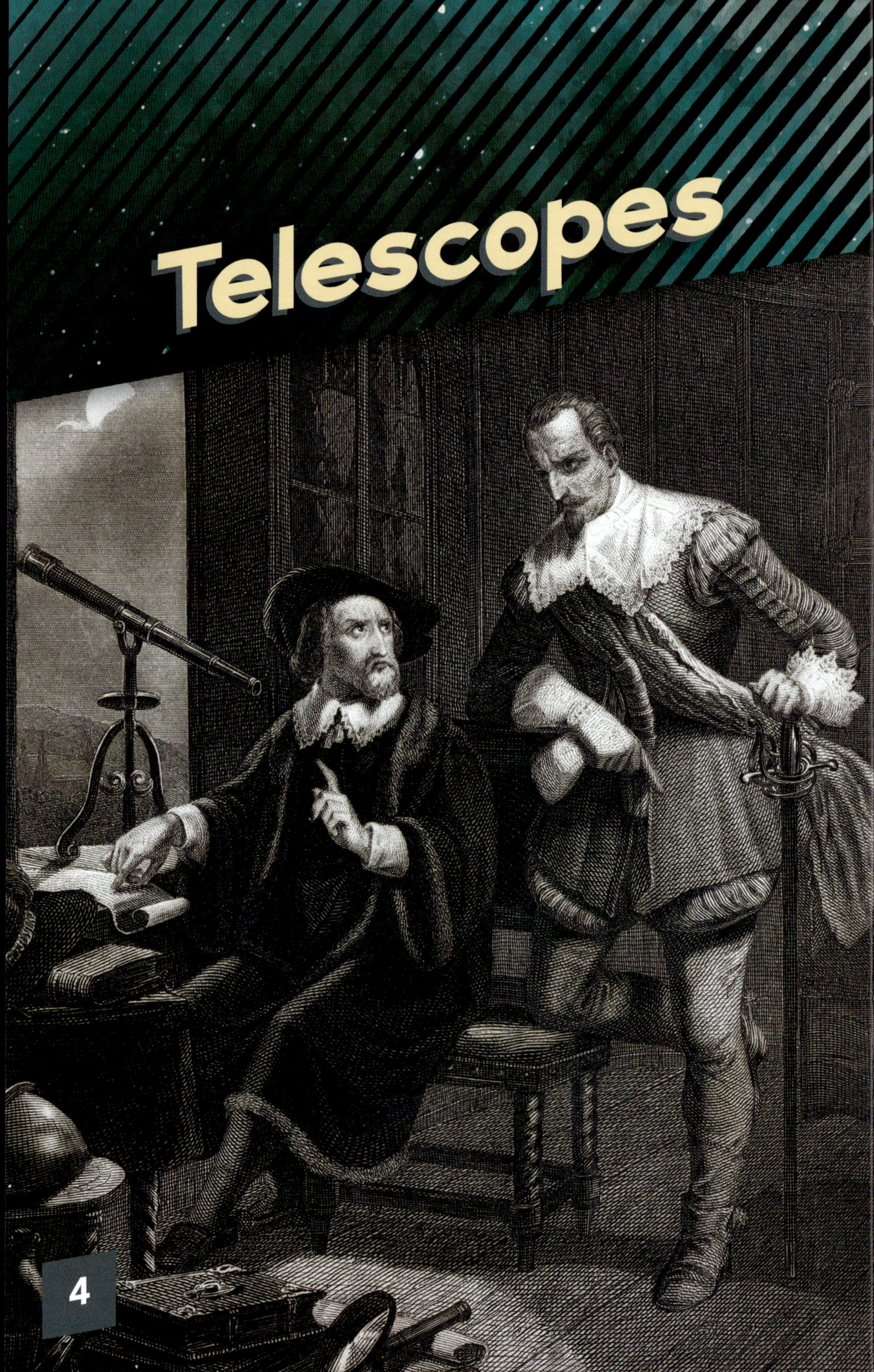

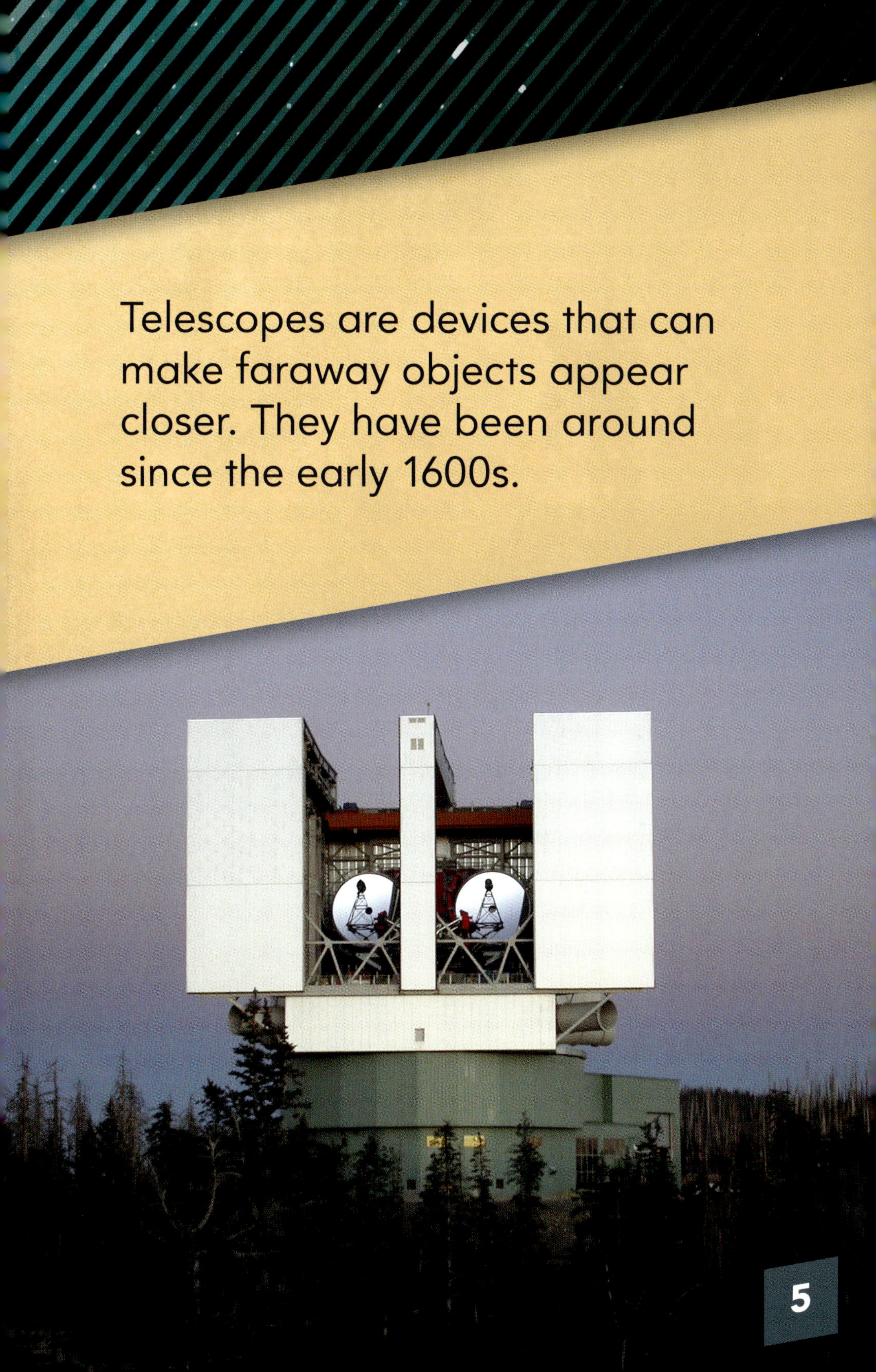

Telescopes are devices that can make faraway objects appear closer. They have been around since the early 1600s.

Today, telescopes can be small ones you use at your house. Or they can be giant ones used to study things very far away. Some are even used in space!

The first operational space telescope was launched in 1968. It was the Orbiting Astronomical Observatory. It was nicknamed "Stargazer."

Hubble Space Telescope

The Hubble Space Telescope (HST) was launched in 1990. It went up with the space shuttle *Discovery*. It was meant to last 15 years. It is still working today!

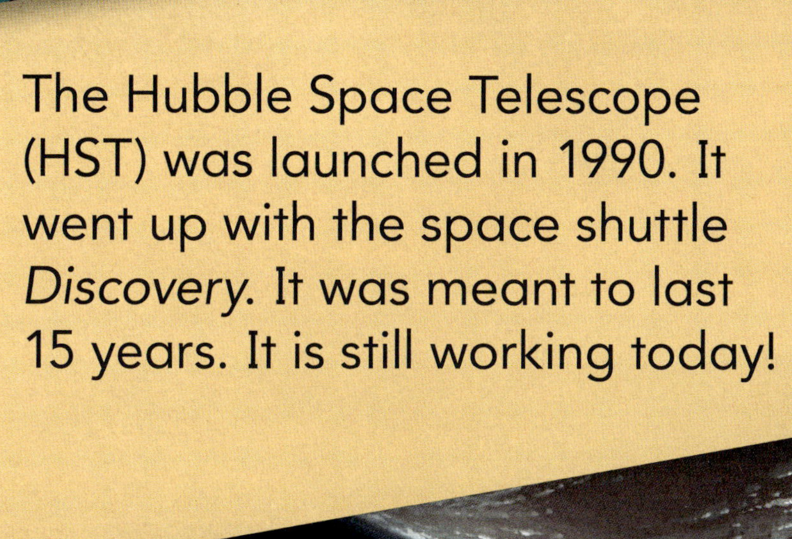

HST has sent more than one million images back to Earth. It has captured stars being born. It has also tracked distant **galaxies**.

HST has been able to study the outer planets and moons. It has also observed giant **black holes**.

Other Space Telescopes

The National Aeronautics and Space Administration (NASA) has sent other telescopes into space. The Compton Gamma Ray Observatory launched in 1991. It detected more than 2,600 **gamma ray bursts**.

The Chandra X-ray Observatory went into orbit in 1999. It can see things that are billions of **light-years** away. The Spitzer Space Telescope went up in 2003. It has looked into the centers of other **galaxies**.

Space telescopes are important tools. They have given us **insight** into how the universe is changing. Future telescopes are sure to provide more exciting discoveries!

More Facts

- HST orbits about 340 miles (547 km) above Earth.

- HST travels 17,000 mph (27,359 kph). That means it moves five miles (8 km) per second!

- The James Webb Space Telescope is scheduled to launch in 2021. Scientists hope it will provide information about how the universe began.

Glossary

black hole – a region or body in space with gravity so strong that neither light nor matter can escape it. Scientists believe that black holes form when very large stars collapse.

galaxy – a collection of billions of stars and other matter held together by gravity. Our planet Earth and the sun belong to the Milky Way galaxy. They are only tiny parts of this galaxy.

gamma ray burst – an extremely energetic explosion that has been observed by scientists in distant galaxies. A gamma ray is a type of electromagnetic radiation of high energy and frequency.

insight – the power to understand deep meanings or truths.

light-year – a unit of distance equal to the distance light can travel in one year, which is about 6 trillion miles.

Index

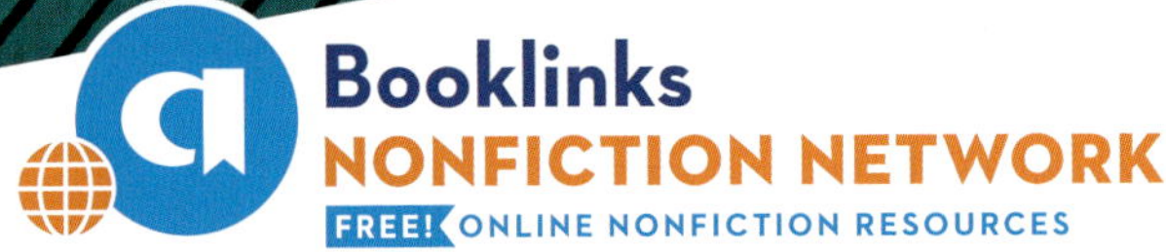

Online Resources